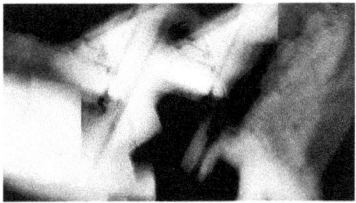

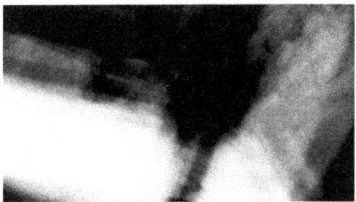

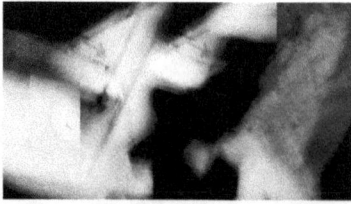

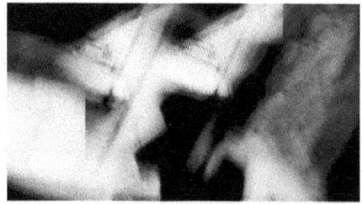

kathy mctavish

night train / blue window

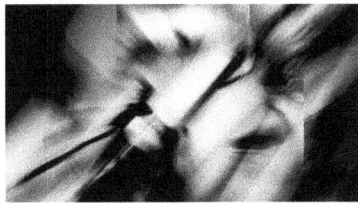

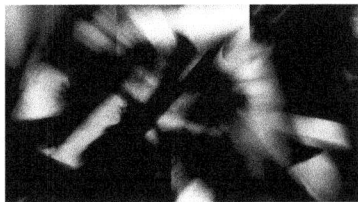

night train / blue window
© 2012 Kathy McTavish

Wildwood River Press
Duluth, MN
www.wildwoodriver.com

Printed in the United States
ISBN: 978-0-9843777-6-3
Library of Congress Control Number: 2012937239

for the planet & in memorium:
1914 / the last passenger pigeon

thank you:
American Composers Forum
Cheswatyr Music Commissioning Fund
Jerome Foundation
United States Artists
Beverly Berntson
Carla Blumberg & Barb Neubert
Hanne-Grete Brink
James Holdman
Peggy Horgan
Jack Kritzer
Michelle Lebeau
Jeanne Maki
Don & Janet McTavish
Karen McTavish
Pamela Mittlefehldt
Erika Mock
Marc Swoon Bildos Neys
Sheila Packa
Tineke Ritmeester
Sharon Rogers
Mary Seabloom
Christopher Shillock
Kent Schul
Ann Tash
Helen Vitoria
Lee Zimmerman

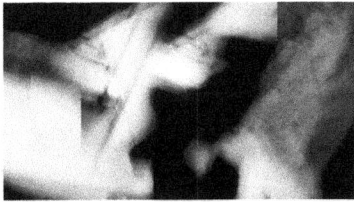

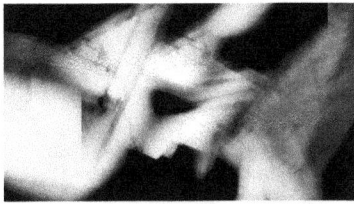

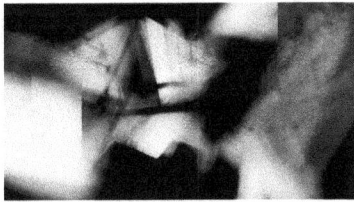

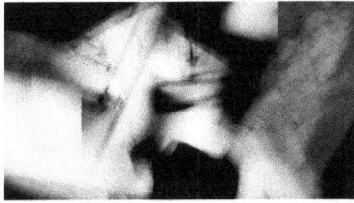

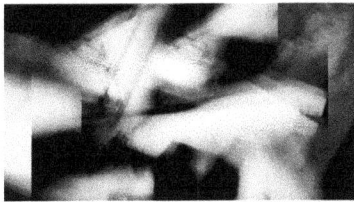

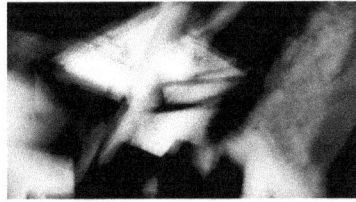

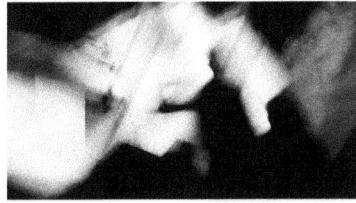

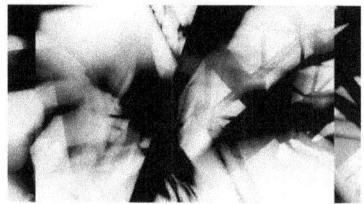
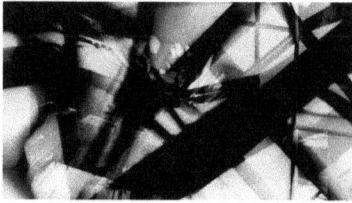

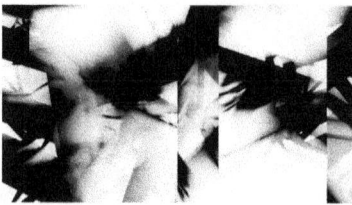

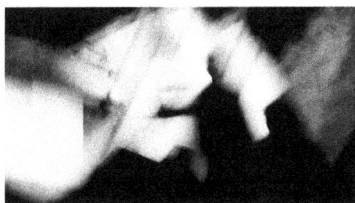

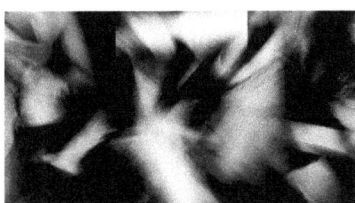

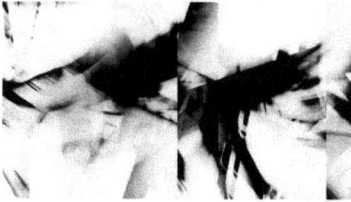

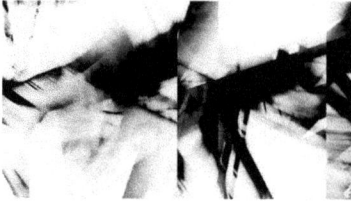

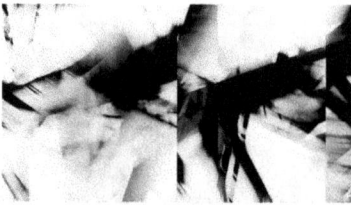

1914 / departures
last stop / cincinnati zoo

martha
died
one
september
day
last
passenger
last
bird

decades before there were
clouds of birds
tunnels of birds
forests ablaze with birds

then
one

then
none

swept by
a silver scythe
an empty swing
a wild &
twisted
flight

last
passenger
pigeon

she
slipped
away
forever
into sky
through
iron bars
into blue
ghost wings
wing bone
tiny xylophone
weightless
kite

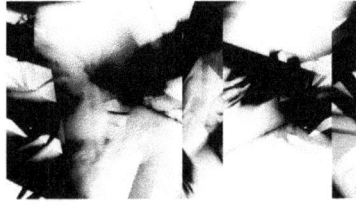

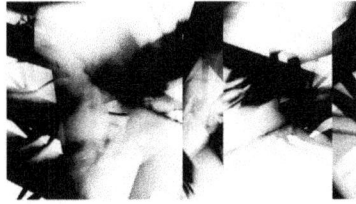

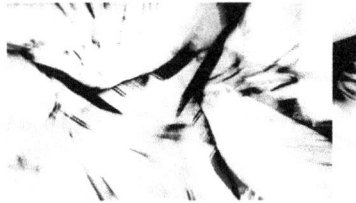

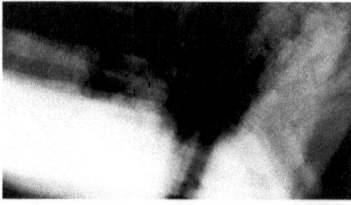

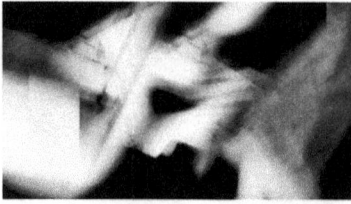

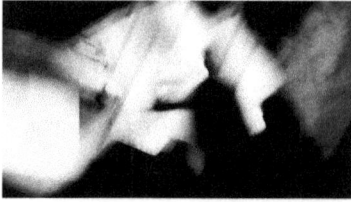

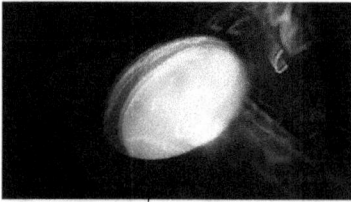
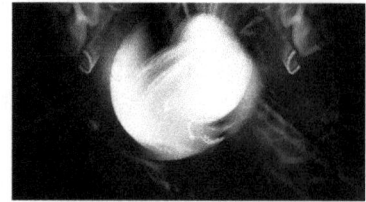
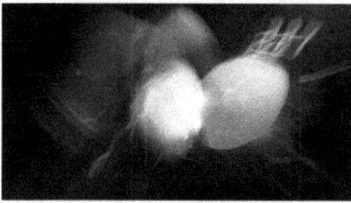
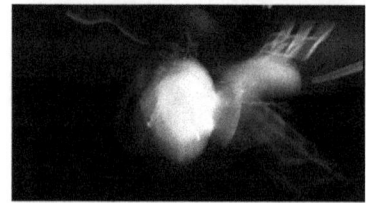

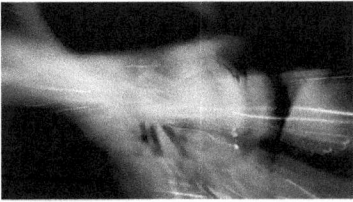
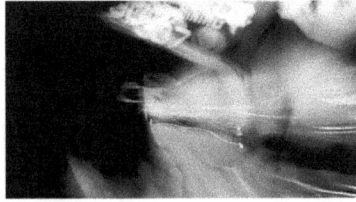
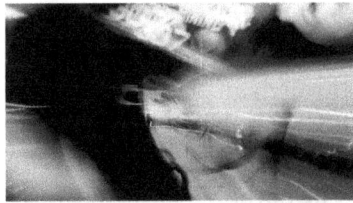

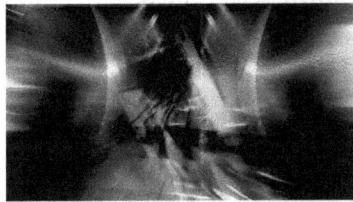
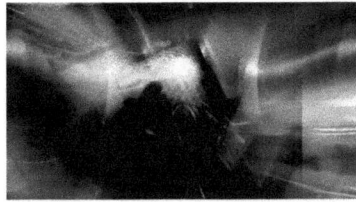
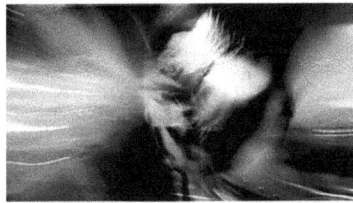

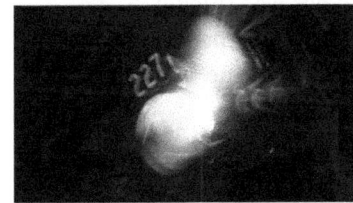
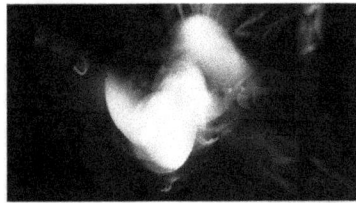
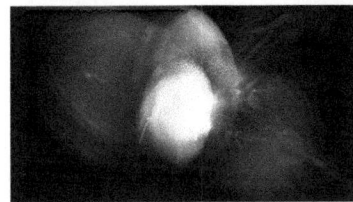
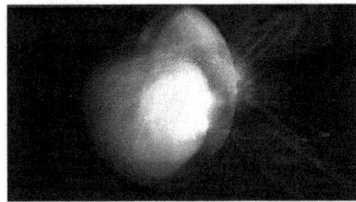

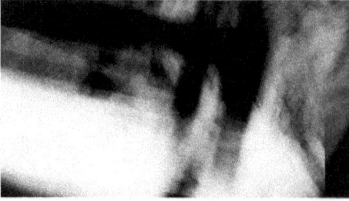

pullman flew

through an
open window into a
blue pool of light
s/he harbored at the nearest
railroad station &
took the night train
hot summer 1914
restless humidity /
season of rain

a bird could easily find
passage
it was just a matter of
unfolding a map
worn by dust & decades
creased in a junkyard chevy
tomb
glass was thicker
roads wound slower &
iron had more
weight
mail made its way from
town to town

14

sun soaked the landscape
rains fell against
tin dark nights
city lamps were
dimmer &
shadows
were deeper
a wild electricity
slipped silver across
naked wires
static expectation
the horizon bustled with
gleaming factories

forget the bullets
shouting towards heaven
& black clouds of wings
raining death
forget about orange fire
ravaging hundreds of forest
miles
thank the ax for
rows of houses
saluting the day

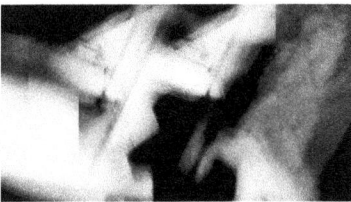

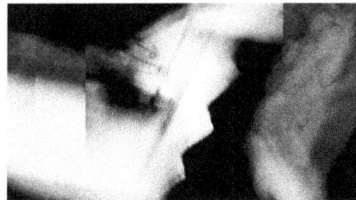

15

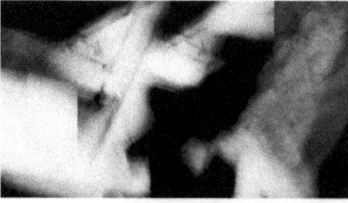

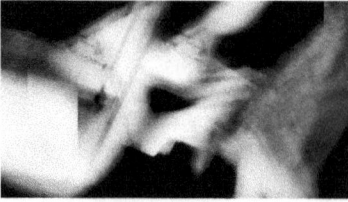

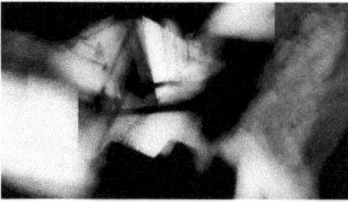

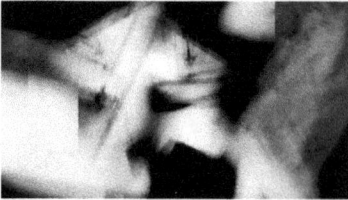

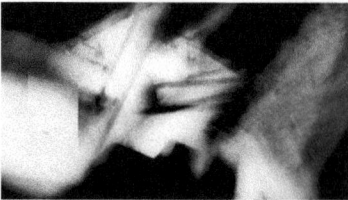

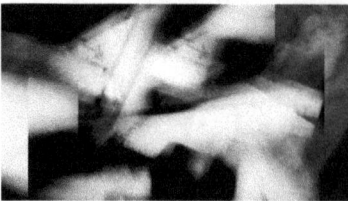

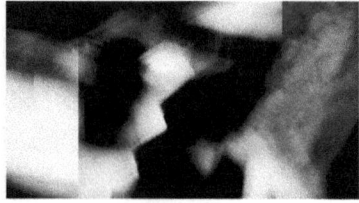

there is a blue window
on this train
one map one clock
one lamp one bird
a passenger named pullman
clutching a worn brown
suitcase with claw & wing
can you hear the lonesome
whistle? call of wind
mouth harp & gentle beast
friend of night
coal whisky grain
homeless ghost
wandering shadow
on this train to glory
one car carries arms
one car carries soldiers
one car carries coal
one car carries grain
one car carries
... look away dear reader
one car carries live stock
brown frightened eyes
look away once more
out the window

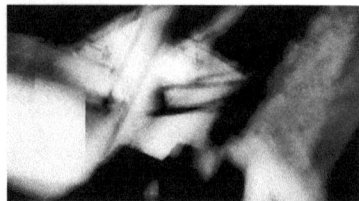

fix your eyes upon the tracks
silver shining
ribbons of moon
the sun bends her rays
around this
rough cut orbit
glancing off our moon (our
only moon)
to dimly stain the night
even now the sun
reaches us with
her lamp her veils
her tenderness glimmers
before morning arrives
to this map this dawn

the mail train
the london, midland, scottish
railway
or iowa nebraska chicago
new york

our train moves forward
destinations west or south

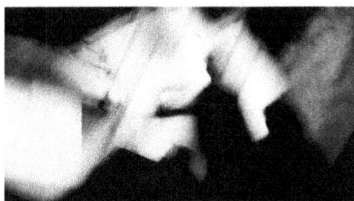

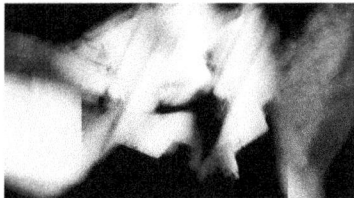

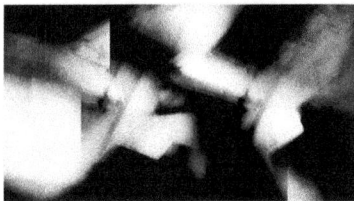

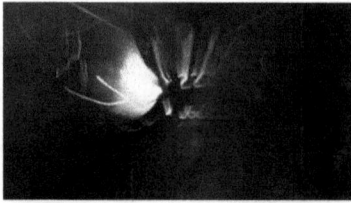

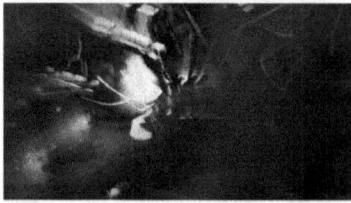

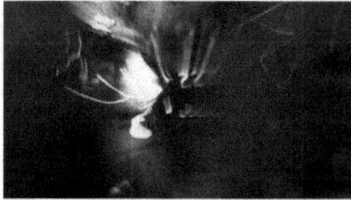

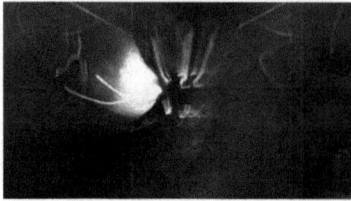

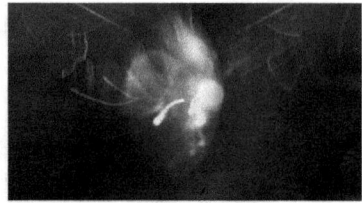

the black glass window
flashes silver
time /
a map

wings & beak
fur, muscle, & hooves
from icy cliffs &
shifting tides
losing you
slipping grip
from this blue globe
this broken circle
fins & bulk
whale or reef
grasslands gone
forests torched & split

falling
a bird but fallen
broken suitcase
broken bottles
reeds for lips
night train hurtling
forward

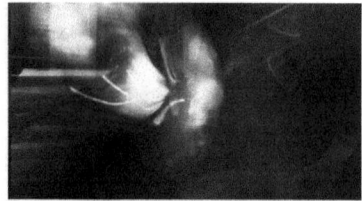

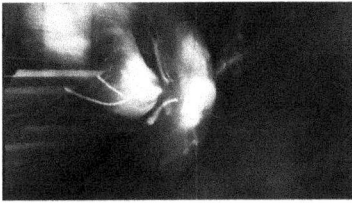

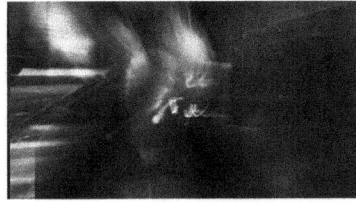

cincinnati / ohio
the blue zoo
last stop
last motel
last call
last station
on the silver rail line
texaco dreams
this is the stop
this is the time
five till midnight
lamps aglow
the bell tolls
the oily pools
parched deserts
silent oceans
birdless skies
& your red eyes
brown
suitcase
torn shoes
coughing bird
whisky bird

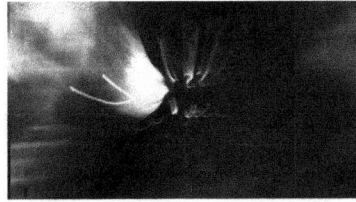

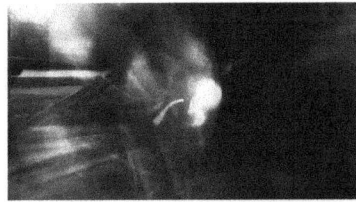

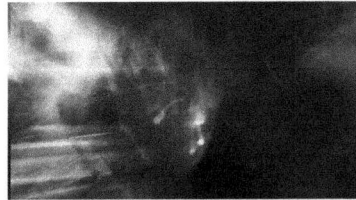

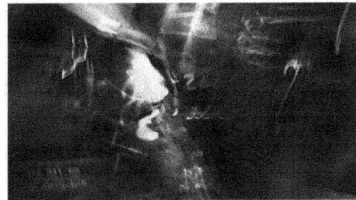

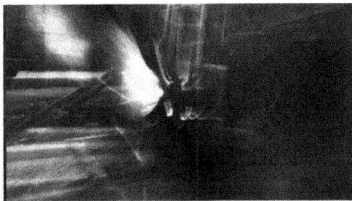

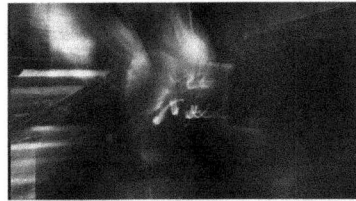

pullman woke

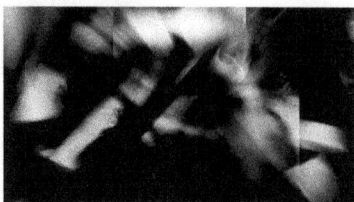

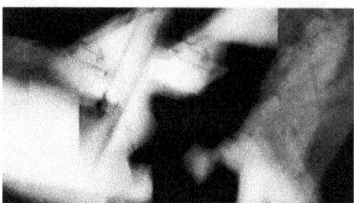

not september
not cincinnati
not caged
not iron bars or
black black oily seas
forests tall as canyons
not september
not that passenger
alone
in her cage
singing to
late night
hollow echo concrete echo

this iron horse rides
past carnivals &
bowling alleys
texaco signs &
bars
industry!
past sleepy towns & drive in
pictures
life aglow
shiny metal coins
& watches

books of green stamps &
toasters
silverware &
pretty red glasses
green rimmed plates
pretty circles
pretty lines
& diners
the horse pulled in
to rest
& pullman found a table at a
diner near the station
near the five & dime
near woolworth's pretty bell
the door opened on bells
a sandwich &
steaming coffee
a soda &
piece of pie
red red cherry
pie
& glass windows
christmas lights & dolls
not december
but august afternoon

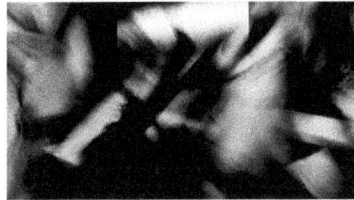

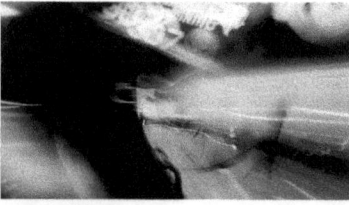

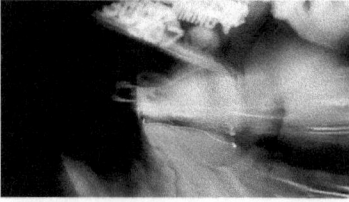

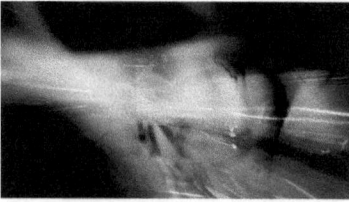

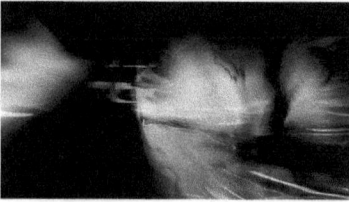

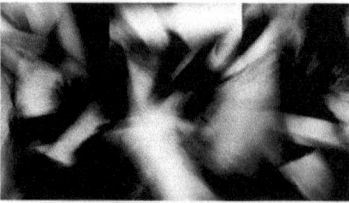

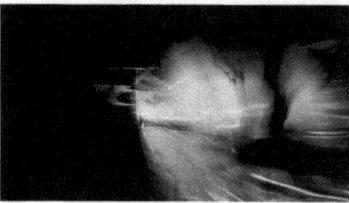

lazy sun
sweltering noon day
sun & glass windows
gathering heat
glory heat
asphalt heat
& pullman
at the center
of the reflection
in radiant heat
the center of all
light
a mirror
a pocket full of mirrors
to hold the heat
black heat
mercury rising
pullman rising
blazing opal eyes
cloaked in radiance
soaked in
radiance
fire
rushing down streets
at dawn

at midnight
pullman gathers his
blue coat
& flies
all aboard
no sandwich
hurry aboard
swing up onto
the platform
iron chariot
rolling fire
wheels & fire
not september
not
a soot lined window
opening onto
cincinnati streets
august hot night
bisected night
maps, trains, rural routes
dust & stones
sky
black lines
night
lamp's yellow

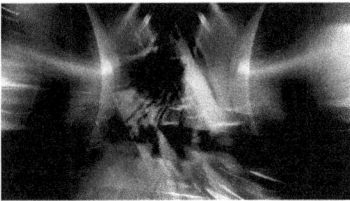

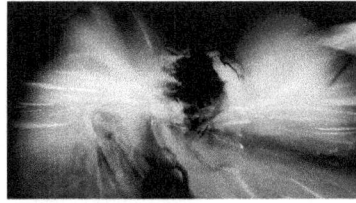

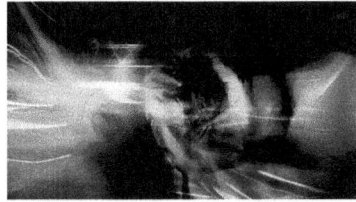

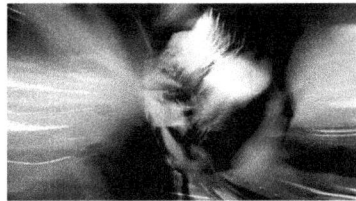

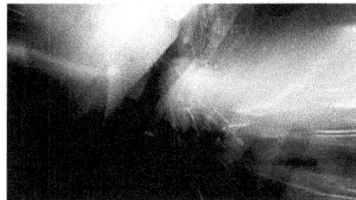

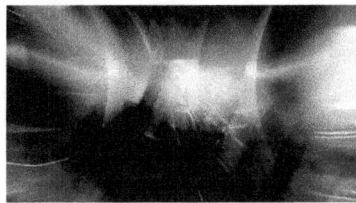

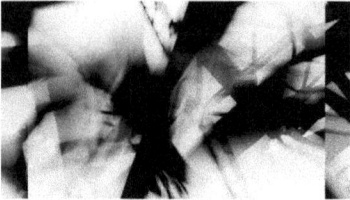

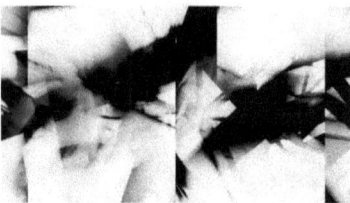

halo &
moth
pullman awake
steam & iron
iron velocity
iron night
steam red fires
black coal
red hat
red ticket
torn
sandwich
pale paper
pale stare
pale conductor
dreaming
with sandwiches
asleep in
broken down
streetlamps
broken light
cold nights
sandwiches
& dumpsters
manifestos & blankets

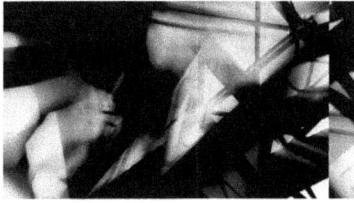

stolen maps
brown paper bags
& white receipts holding
black rows of numbers
blue patches of light
loose change &
crumpled bills or
gleaming mountains of green
golden houses built on
acres of green
rivers of black
asphalt between
a bird
would fly far from here
but pullman was
a traveler
a passenger
a resident of this iron box
this iron speed
a passenger of forward
motion
forward eyes
hurtling motion
next station
next silver dawn

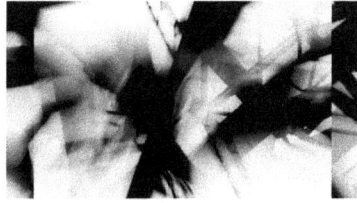

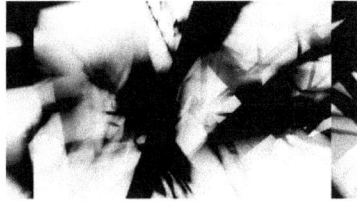

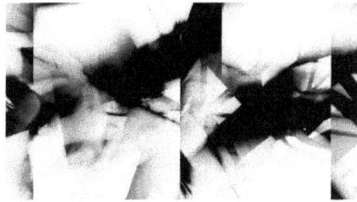

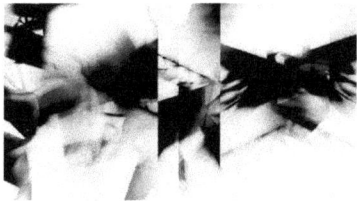

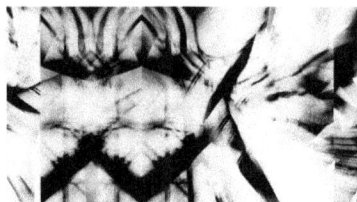

next sun twisting
planet bending
light
or hOwl
tight beaked
glowing eyes
relentless forward
iron momentum
a book
on gravity
a window
down
it was the plan
of
pullman
to pull it down
& by downward falling
to turn the motion
backward
to that day
to introduce
a window
towards sky
slate heavens
& passage

a 3rd possibility
a forever
august
a large round
still clock
in ecstatic suspension
arms flung wide
across the golden
sun
shadow angels
falling
hYmns
a table of sandwiches
red drinks
& steaming
cups of
coffee
vending machines
sky blue
& nickels
spilling across the floor
pullman could see
this 3rd possibility
a vision or
formula

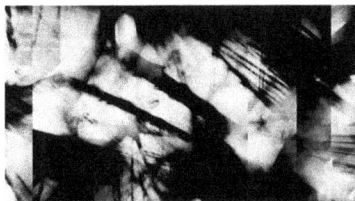

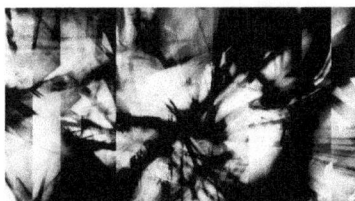

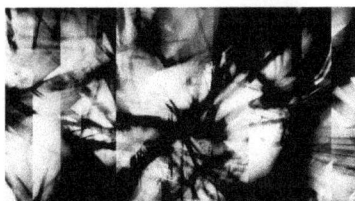

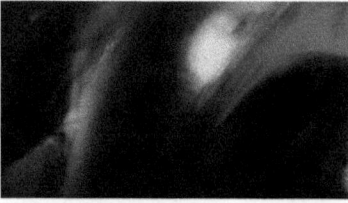

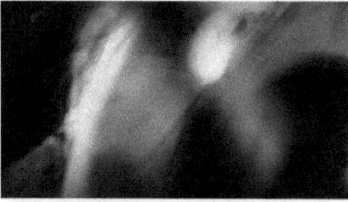

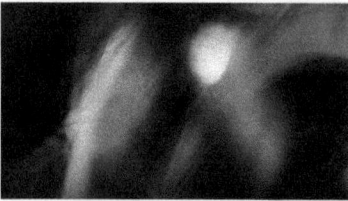

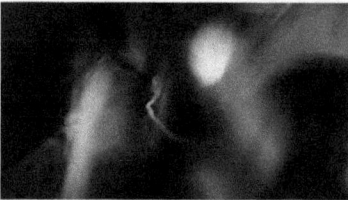

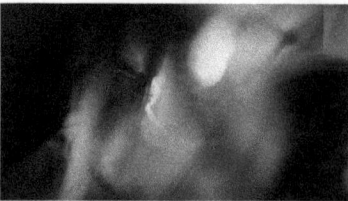

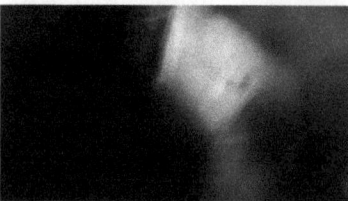

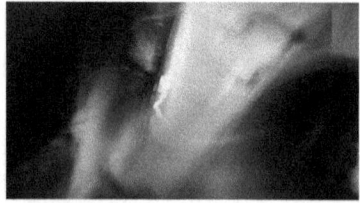

in chalk on
slate
or bound
in leather
books
illuminated
red & gold
black black ink
an angel
falling backward
bisecting time
lancing wrong
turns
stitching extinctions
back to fabrics
replacing maps
back towards before
that day of
shadow birds
a pen lit
drawing
the train
a memory of metal
iron black
lonely singer

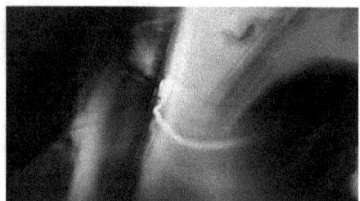

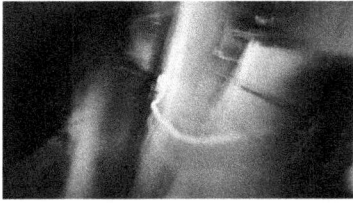

night & red & coal
velocity & weight
relentless hammer
chest
& weight
eyes & pitch
black ink & rain
twisting tracks
racing mass
gleaming night
lamp
stealing past
yellow silos
grain & coal
or lonely eyes
gaunt rider
wool fraying against
the night
the dirt dark floor
the track & heat
a glow
an iron beast
a roar
here on this old
platform

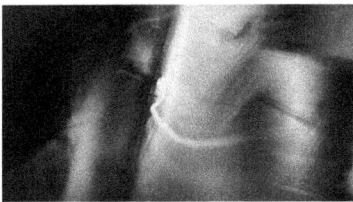

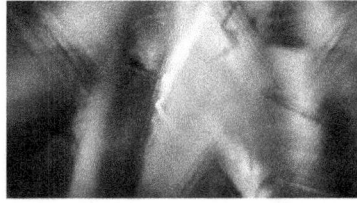

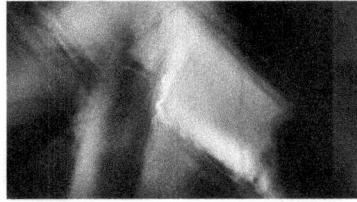

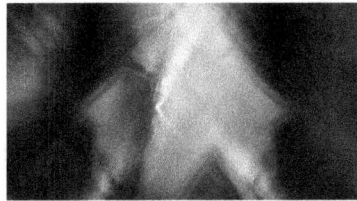

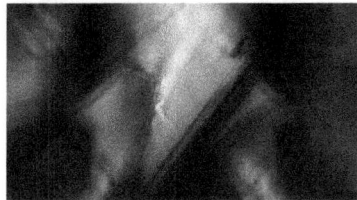

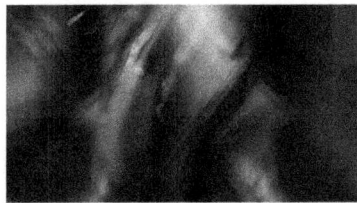

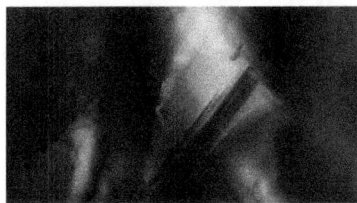

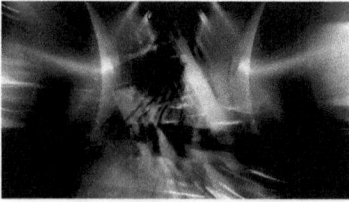

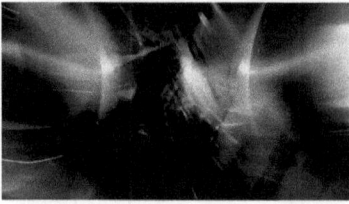

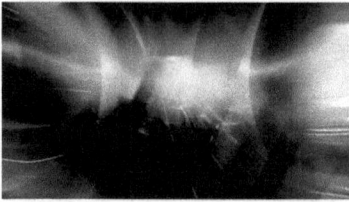

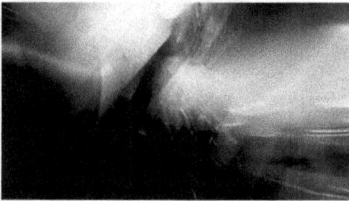

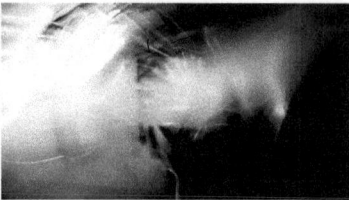

left behind
on splintered
floor dark & lonely
lampless floor
on a platform a
bird clutches a
suitcase with broken
clasp
dusty lungs
taste of far away
dusty sky
or the train
unbound
in its belly
gaunt figure
dark crow
night blanket
relentless motion
black steed
train of night
train of iron
train of fire
the horse the grain
the coal the fire
the dawn

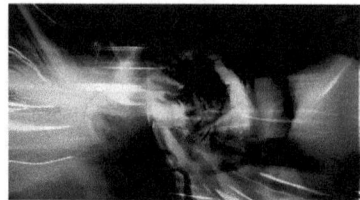

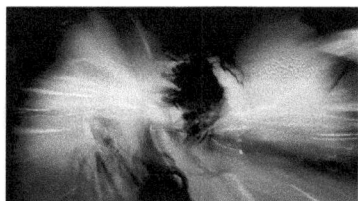

unstoppable unmerciful
unending velocity
blue iron
mass & holy speed
night rider
grain
commerce
palettes of steel
lumber broken trees
yellow pupils
black cavern
red flames
thunder
god of storms
silver messenger
god of fire
welder of earth & sky
electric sky
the sky was falling
in glassy shards of
blue & steel
soot & rain
cold night
winter of atoms
wings turned to vapor

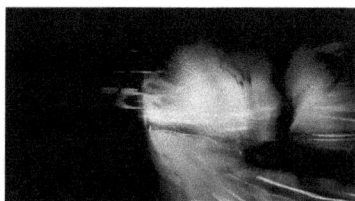

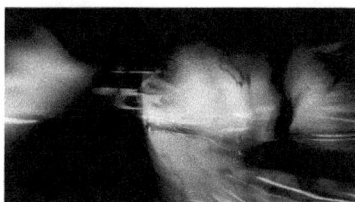

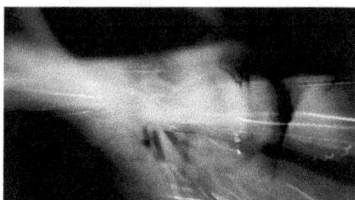

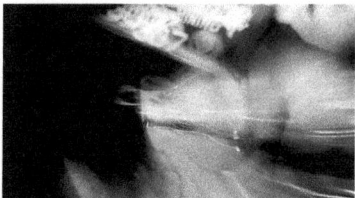

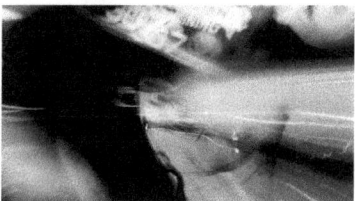

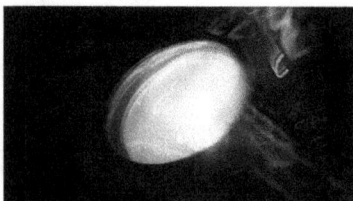

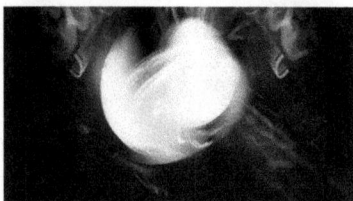

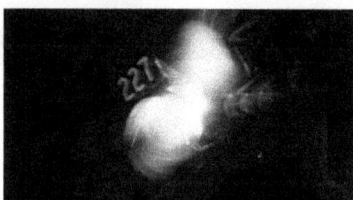

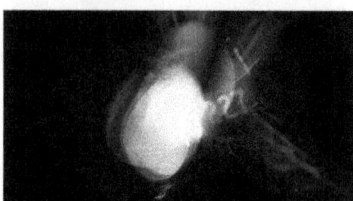

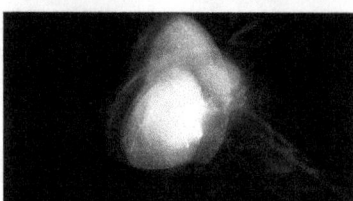

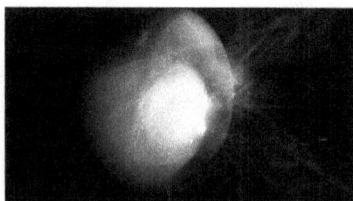

hearts turned to stone
desert eyes
shadow humans
chalky outlines
burned into concrete
formulas lining books
inky extinctions
the whimper of gone
last breath
downward curve
a population dynamic
& finally last bird
last flight
last wings
last breath
hYmnless night
shadowless sun
merciless trajectories
heartless calculations
empty cage
rip tide & storm
or ghostly hollow noon
still noon
arid noon
white heat &

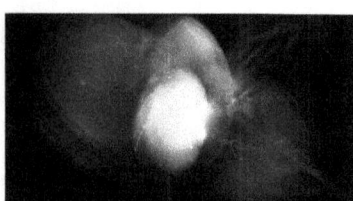

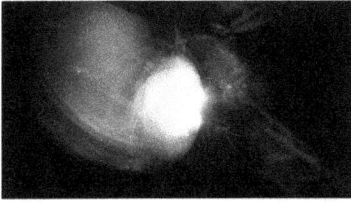

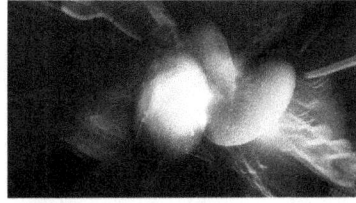

absence
a hole in the sky
small planet
small circle
oily seas
coal clouds
last transmissions
majestic refineries
nevada plumes
atomic silos
jet streams
metallic giants
gleaming glass
white lines
asphalt tides
but long before that
pullman alone
& night train
iron tracks
armaments & cigarettes
voyage of goods
orange agents of fire
transport commerce
iron waves
relentless tides

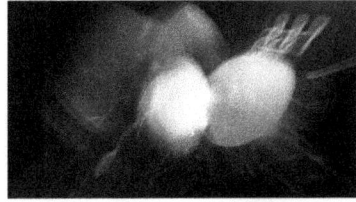

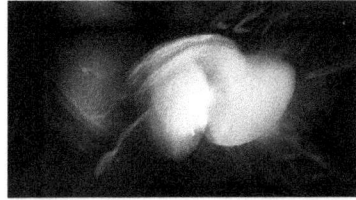

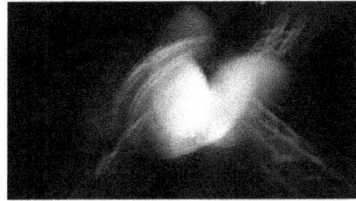

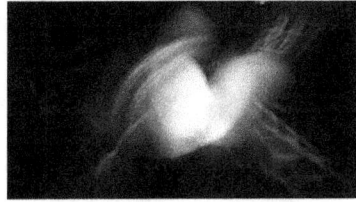

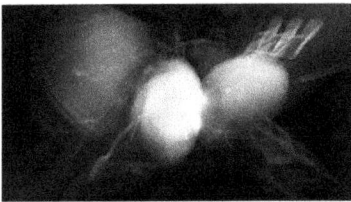

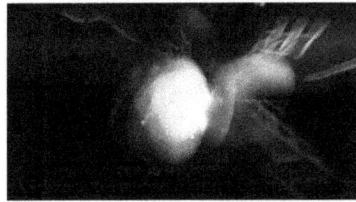

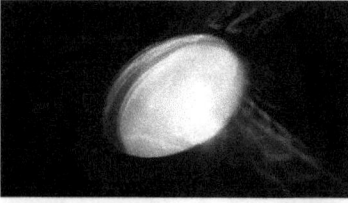

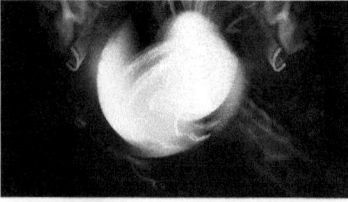

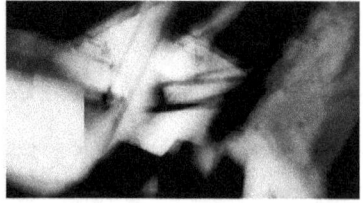

rising
body counts
rapid heartbeat
extinction's pulse
trashcan symphonies
vacancy
empty lots
or endless
high rise
concrete &
broken glass
twisted steel
dynamite cyclones
but long before that
pullman &
different maps
different trains
faraway night
last stand
texaco
paths
chance
infinite between & time
light harp in 4 walls
light hYmns

light color pitch
sound
depth or time
plastic commerce
& neon bibles
litanies
sacrifice & risk
broken hYmns
near sighted crow
turn the clock
bend the metal
throw the switch
slam the lever
traversals
migrations
no line falling
digits in flight
letters numbers
electrical wires
radio waves
wireless traffic
electron rush
epic flying movies
traveling through air
invisible waves

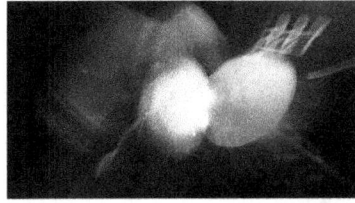

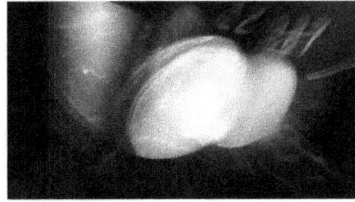

people talking
through air
heavy clouds
information super highways
electron prophets
but long before
silicon chips
the race of numbers
pullman's chipped black beak
tattered red shirt
blackened boots
& worn brown suitcase
brown parcel
traveling home
another map
creased & stained
other trains
other nights
lonely crowds
guards on duty
sentinels
barbed wire
& prison gates
but before elevator
resurrections

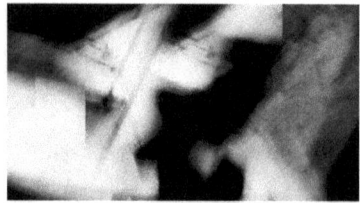

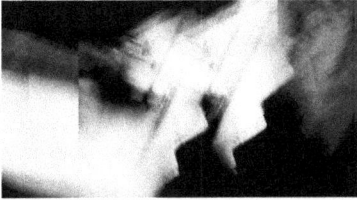

asphalt meadows
atomic roulette
& broken webs
shattered webs
before guns
last retort
pullman riding
riding home
or last bird
dreaming
blue blue window
floods & sirens
night rider
last dream
& her

the moth
captive of fire

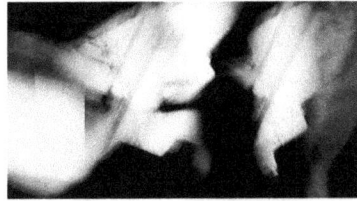

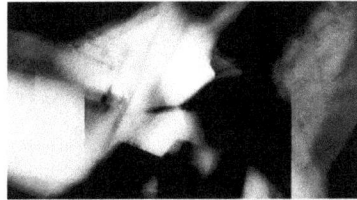

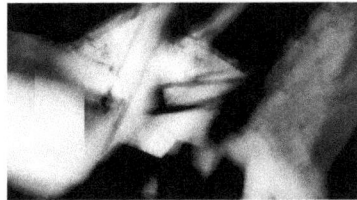

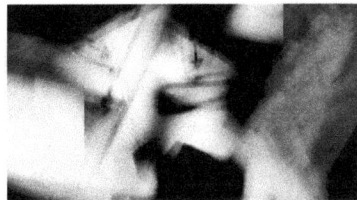

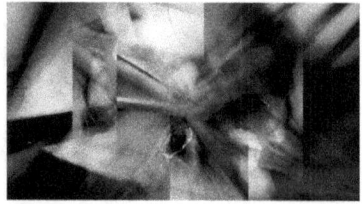

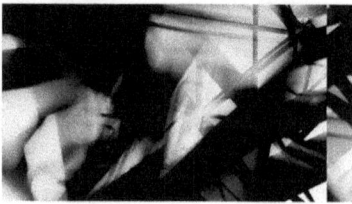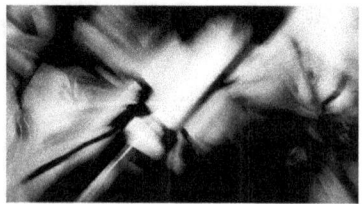

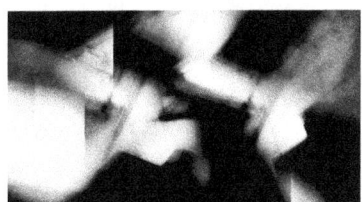

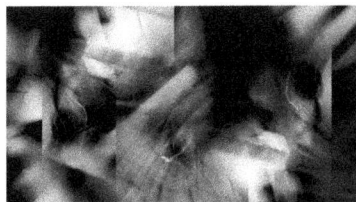

thewheeltheflamethewheeltheflamethewheelthe–
flamethewheeltheflamethewheeltheflamethewheelthe–
flamethewheeltheflamethewheeltheflamethewheelthe–
flamethewheeltheflamethewheeltheflamethewheelthe–
flameholyholyholyholy**holyholyholy**holytheflamethe–
wheeltheflamethewheeltheflamethewheeltheflamethe–
wheeltheflamethewheeltheflamethewheeltheflamethe–
wheeltheflamethewheeltheflamethewheeltheflamethe–
wheeltheflamethewheeltheflamethewheeltheflamethe–

40

wheeltheflamethewheeltheflamethewheeltheflamethe—
wheeltheflamethewheeltheflamethewheeltheflamethe—
wheeltheflamethe**thewheel**theflamethewheeltheflamethe—
wheeltheflamethewheeltheflamethewheeltheflamethe—
wheeltheflamethewheeltheflamethewheeltheflamethe—
wheeltheflamethewheeltheflamethewheeltheflamethe—
wheeltheflameholyholyholyholyholyholyholythe—
wheeltheflamethewheeltheflamethewheel**theflame**the—
wheeltheflamethewheeltheflamethewheeltheflamethe—
wheeltheflamethewheeltheflamethewheeltheflameholy

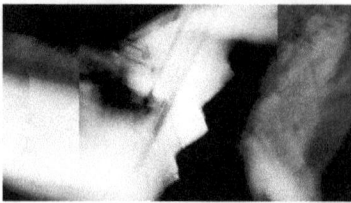

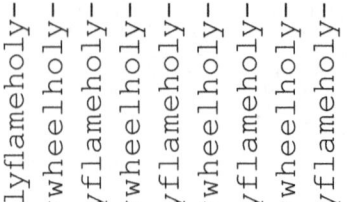

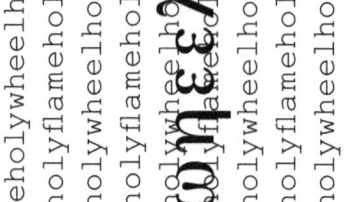

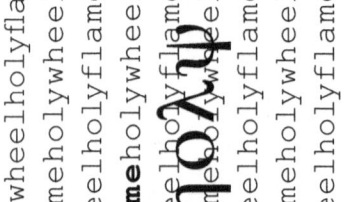

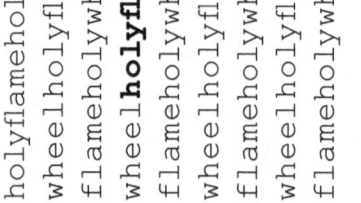

holyflameholywheelholyflameholywheelholyflameholy—
wheelholyflameholywheelholyflameholywheelholy—
flameholywheelholyflameholywheelholyflameholy—
wheel**holyflame**holywheelholyflameholywheelholy—
flameholywheelholyflameholywheelholyflameholy—
wheelholyflameholywheelholyflameholywheelholy—
flameholywheelholyflameholywheelholyflameholy—
wheelholyflameholywheelholyflameholywheelholy—
flameholywheelholyflameholywheelholyflameholy—

wheelholyflameholywheelholyflameholywheelholy—
flameholywheelholyflameholywheelholyflameholy—
wheelholyflameholywheelholyflameholywheelholy—
flameholywheelholyflameholywheelholyflameholy—
wheelholyflameholywheelholyflameholywheelholy—
flameholywheelholyflameholywheelholyflameholy—
wheelholyflameholywheelholyflameholywheelholy—
flameholywheelholyflameholywheelholyflameholy—
wheelholyflameholywheelholyflameholywheelholy—
flameholywheelholyflameholywheelholyflameholy—
wheelholyflameholywheelholyflameholywheelholyholy

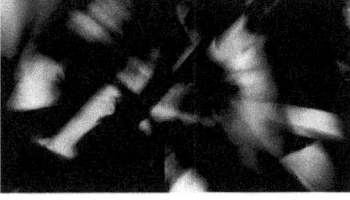

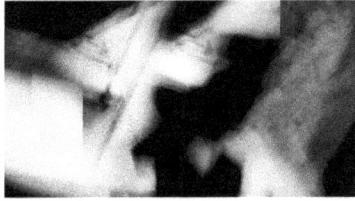

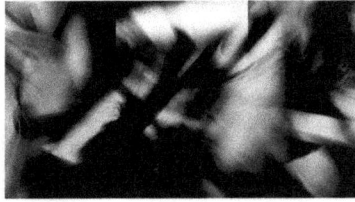

move thyou

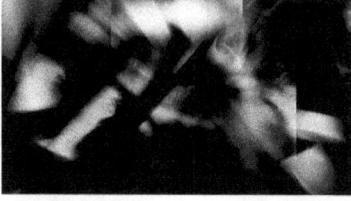

Incipisq uatiunt ipsa cum facerchillab il ius
de et, qui consed modis eum, ut undicae ilicabo
reperorporem doluptas nusdam vitiant inven-
diae ritatiur. Suntotae voluptatem utem eosae
voluptur. Quia voluptatiore hcindus dolo-
riae ius adcaepud qua volubit retis, ende-
bitis modit diae re id moloristia pora cum untio
mostius arum sintota holy wheel etur? Cae dolore
omnis aut ab incto ducianti tecaboribus qui net,

01011100

simusci magnistium ut omni aut adi que sunt audi tes ullabo. Umquas et et excessi omnimust volor res vendiam quam dolorep tatatibus et laborpo repudi dipidendandi in holy flame quid quia venis velit a nent iusa conse is recto doluptas asitat faccuptio verdebitiorum venvent, et frugitas enisque sit maxim reperes doluptatur? Quis molore sima nullitias tust, sequiae mod pidest, volest inctem qui repel maiossequi quam ium, sin ium

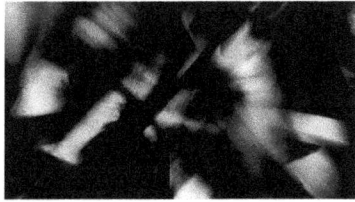

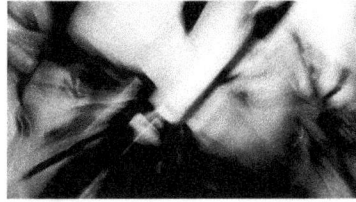

0011010

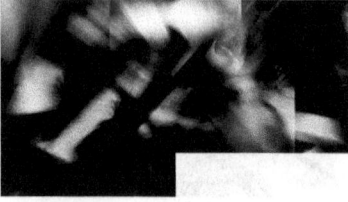

notes:

sunt audi
st volor
laborpo
iia venis
s asitat
fugitas
is molore
, volest
sin ium

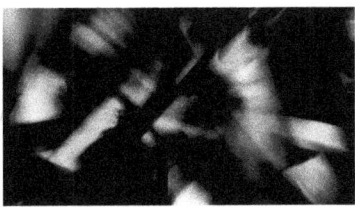

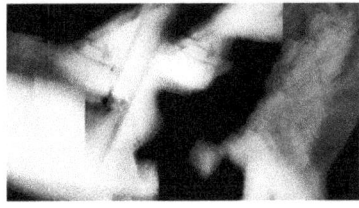

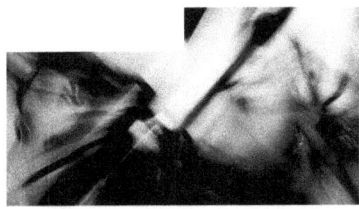

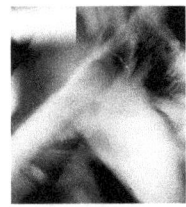

simusci i
tes ulla
res vend
repudi d
velitia i
faccumqu
enisque
sima nul
inctem q

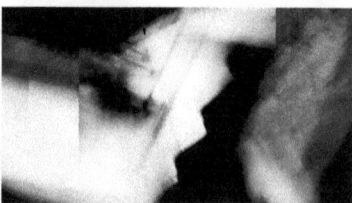

il ius
ilicabo
inven-
em eosae
us dolo-
ss ende-
um untio
e dolore
qui net,

important
dates:

Incipisq
de et, q
reperorp
diae vit
voluptur'
riae ius
bitis mo
mostius
omnis au

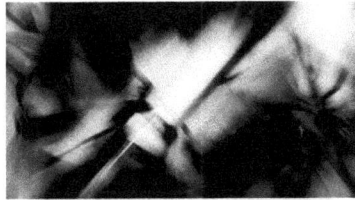

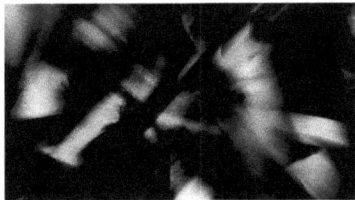

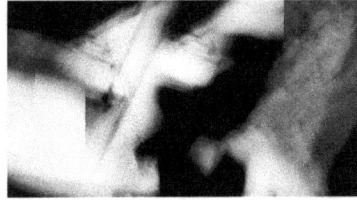

sunt audi
ist volor
laborpo
iia venis
is asitat
fugitas
is molore
, volest
sin ium

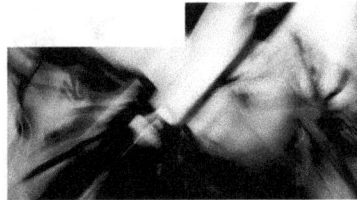

simusci i
tes ulla:
res vend
repudi d
velitia l
faccumqu
enisque :
sima nul
inctem q

1914

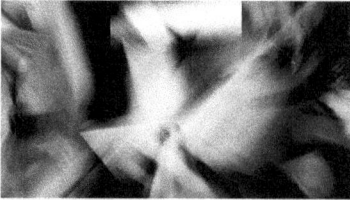

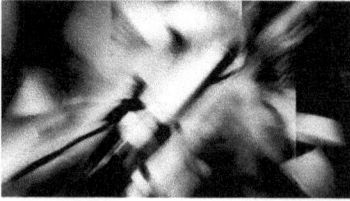

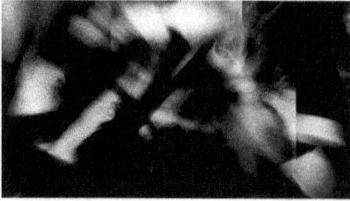

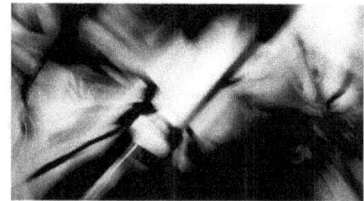

graphite, crayon, charcoal, ragged ink / an arc, jagged, frayed, rounded curve, tonal, bendable, iron, steel string, rope, tone row, clothes line, telephone wire, spray paint / dark alley, sound, light, friction bow & wire, infinite line, bent, frayed, curved, angular, lattice of strings, bone, wood, wire, pine pitch & horse hair, equations & formulas / algebraic manipulations

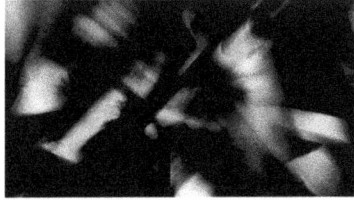

characters: watches for sale, river icarus, pullman, night crow / no time, graffiti angel, river!industry, red stars blue night black wings, the washington avenue bridge, railway icarus / subway icarus .... so many ways to draw a line a story a narrator a solution clues formulas — chalk & slate — a web of interconnected electrical visions — facts sacrificial failures, luminous wings / wax kites — weather & a planet turning

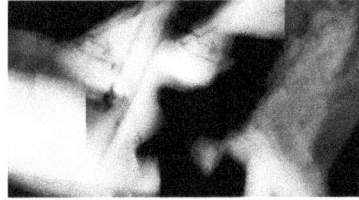

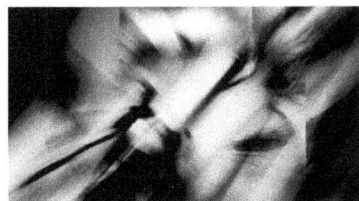

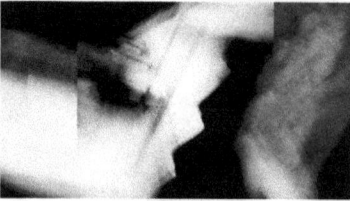

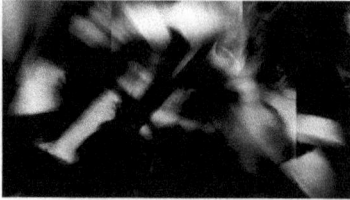

calendars & clocks / church towers (late-night drunken bells) — a time / a place — a train — velocity & physical particles light waves sine waves 1 sun 1 moon 1 blue blue planet a text / the book the Book the word the words the chapters & sections & dates — the *ledger* — count of birds — a glacier a blue blue sea — a window a boat & ishmael's dreams (an ocean heaving) / a fall / a rope & mast ... so many ways to draw

a line ... silver-stringed beast, wooden bird, pine, bone & pale horses ... last departure last sign last bird last stop with captain ahab last train blue window woolworth dreams bells bright august noon sandwich cherry coke gleaming pack- ages shiny shoes new brown suitcase silver clasp a lucky strike the daily news ... industry!

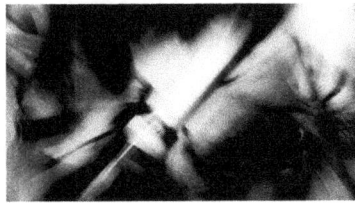

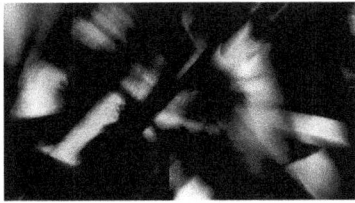

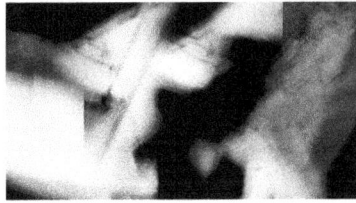

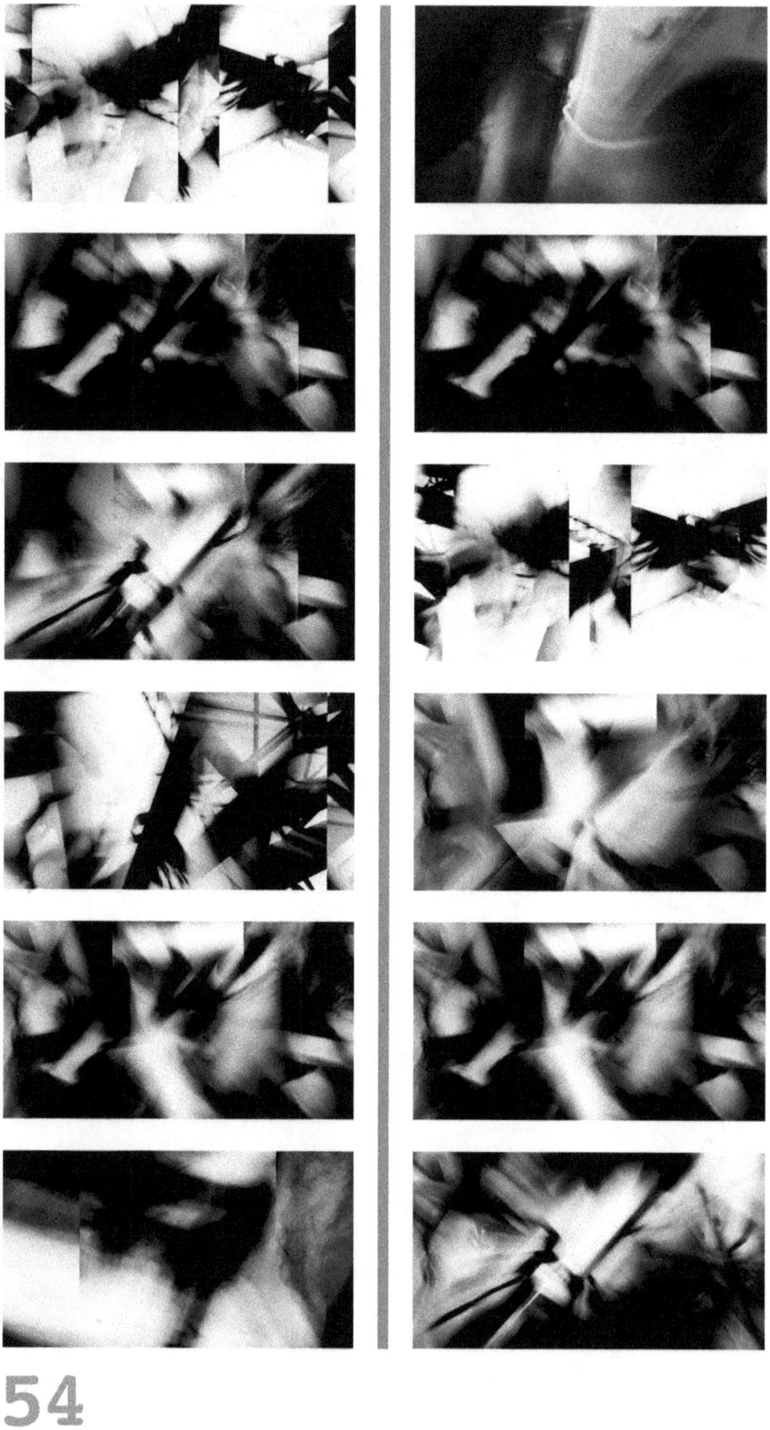

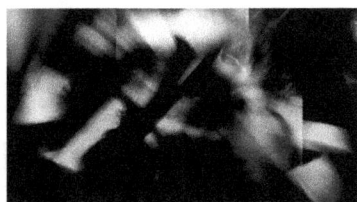

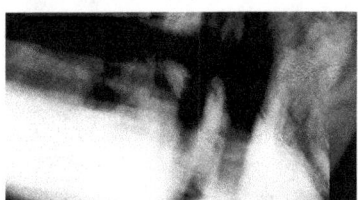

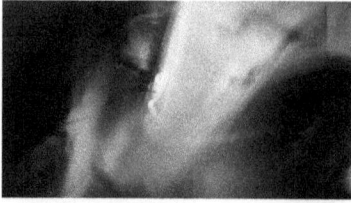

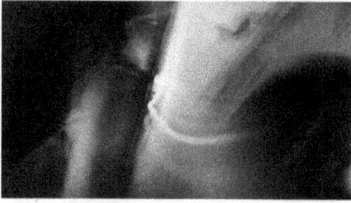

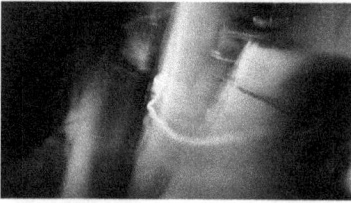

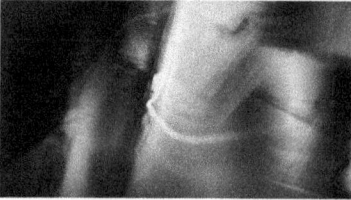

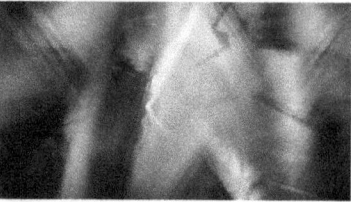

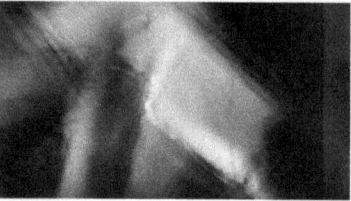

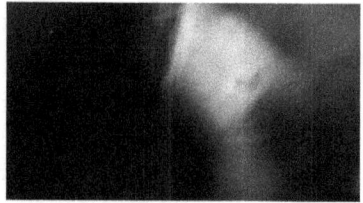

chapter 51 verse 4

incoherent tenuous
broken lines broken light
stormy manifestos
unutterable unseeable
wordless sound
emergent dream
drowning
blue blue windows
manifesto
no map
bendable pitch bendable time
space is time is a path is a
lattice is a train is a cloud
risk & awkward sacrificial
failures
beautiful cranes
language & cranes
transfigured flight
awkward flight
fallen grace
relentless wave
tidal force
extinction / gone

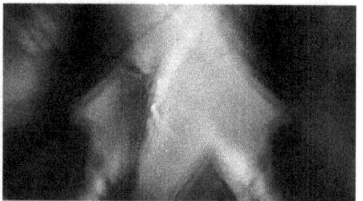

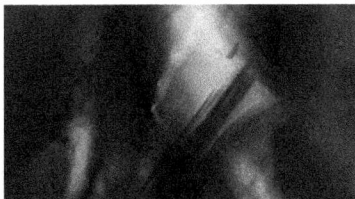

ways to disappear
time
chapter 105 verse 4
flow or flux of time
stream of time
"time's winged chariot"
or train
chapter 111 verse 7
transient transitory impet-
uous fleeting fading flying
mutable fragile
passenger
vagrant wanderer drifter
mariner nomad
bird
flight boat road
so many ways to disappear
transience impermanency
savage formulas
heartless calculations
zero
the
equation
the calculation

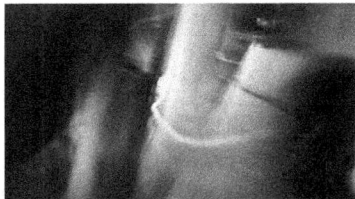

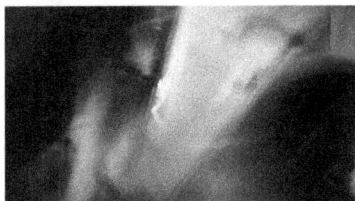

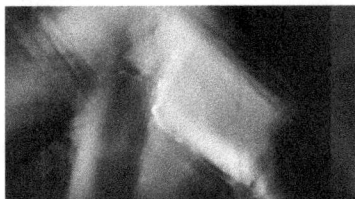

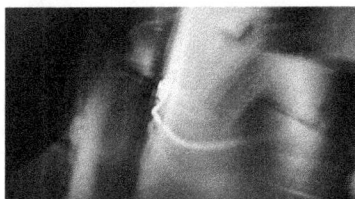

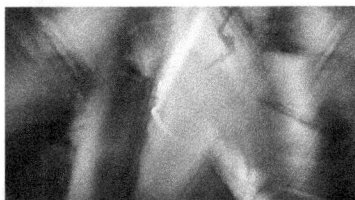

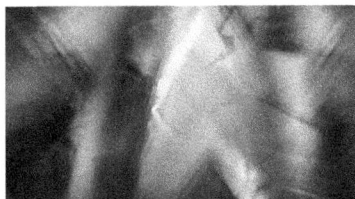

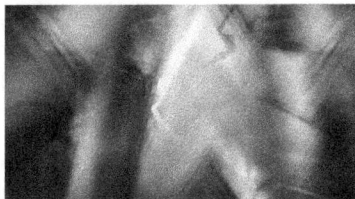

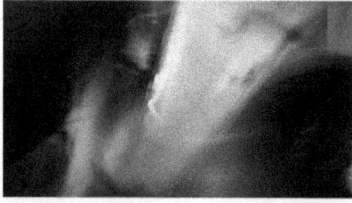

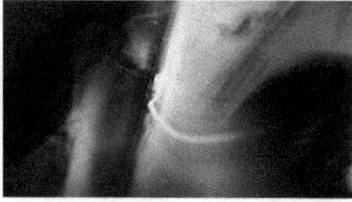

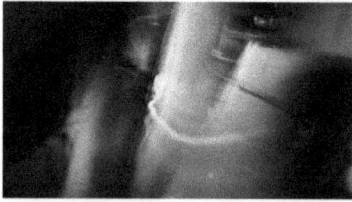

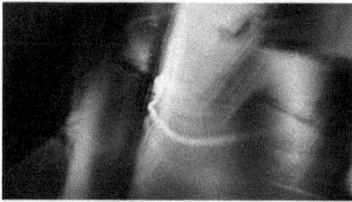

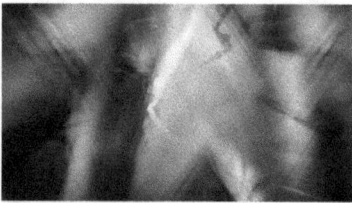

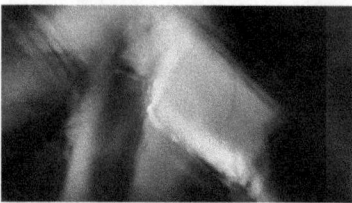

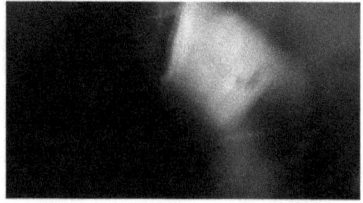

gone

just a drawing on a cave
a blot of ink
a story a filament a history a
photo a grainy recording last
voice last sound last breath
drawer full of
specimens and you
last bird
a name a cage a zoo
a window & pale conductor
every morning greets you
with packaged pellets
barren seeds &

finally
last morning

last entry
september 1, 1914
the last
passenger pigeon died
in the cincinnati zoo

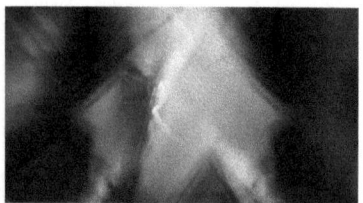

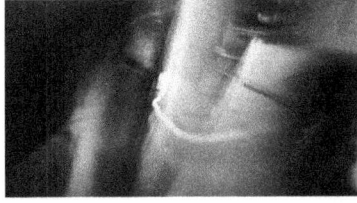
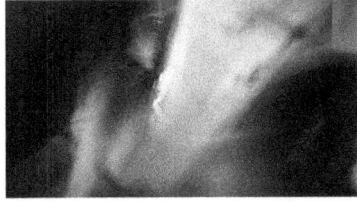
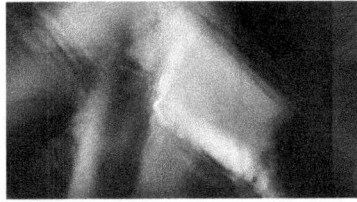
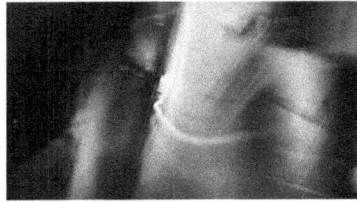
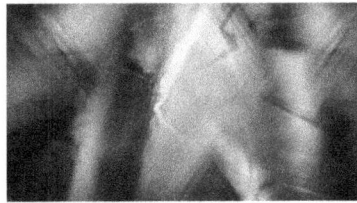
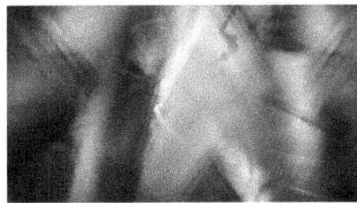

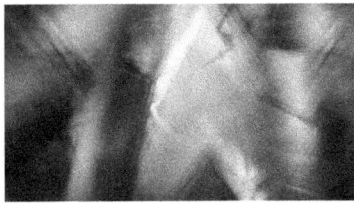

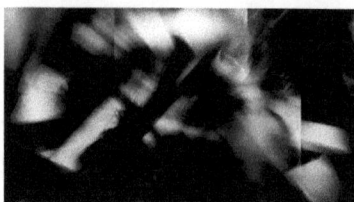

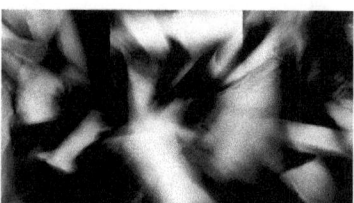
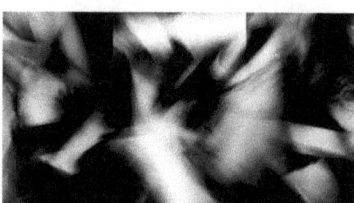
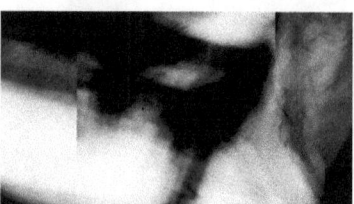

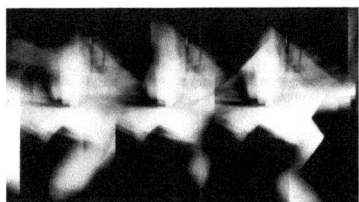

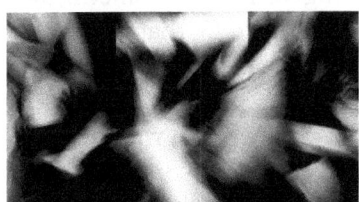

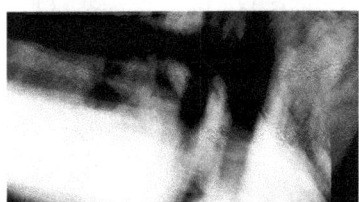

Artist notes:

I am a free-improvisational cellist, composer and multimedia artist. I approach my multimedia work as a synesthetic composer mixing sound, light, texture, color, and lines. My compositional process is like sculpture — blending wood, wire, rosin, horse hair (the cello), found sound, and moving images into a polyphony that blurs the senses.

I have a background in classical cello performance, composition, and mathematical ecology. Ten years ago I began exploring improvisatory musical forms through collaborations in music, theater, film, spoken word, and dance. I started to work primarily as a solo cellist and later began incorporating found sound and my own abstract still-motion video into live performance and recorded work. In the context of my abstract, still-motion projections, the cello takes on the feel of a wandering narrator — sometimes buoyed and sometimes trapped by that frenetic, close, blurred visual environment.

As an ecologist I am moved by the fragility of the planet. Much of my work expresses the broken yet luminous world around us. As both a musician and a mathematician I am fascinated by multi-threaded, dynamical systems and

chance-infused, emergent patterns. As a queer artist I am interested in the ways we construct personal stories / myths and the infinite, bendable between.

There is a companion film for this book also called "night train / blue window." The abstract film and text both center on the ill-fated efforts of a bird named pullman to turn back time to before that day in 1914 when the last passenger pigeon, Martha, died in a cage in the Cincinnati Zoo. This work was made possible through the generous support of those listed in the front of the book.

In 2012 the Jerome Foundation and the Minnesota State Arts Board funded a two-part work called "graffiti angel / holy fool." A 2009 / 2010 commission through the American Composers Forum and the Jerome Foundation produced a work called "river icarus: rusted bridge / deep water." There is a loose band of characters that has emerged from all of these works that I call the "celluloid afterlife collective" after a prose poem by Sheila Packa. Boundaries are blurred between the abstract storylines running through all of these projects. Night train / blue window explores one facet of this larger non-linear, evolving story. I think of this book as a score.

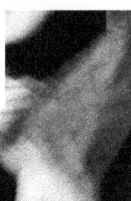

63

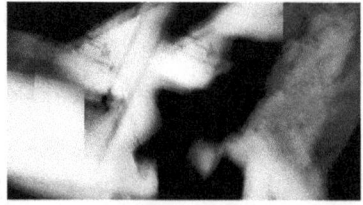
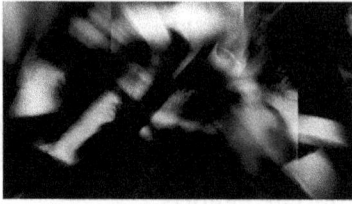

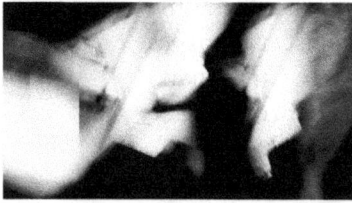

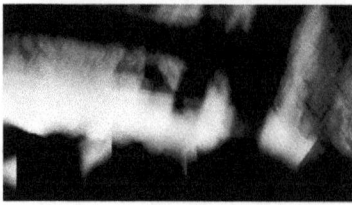

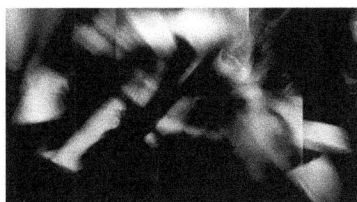

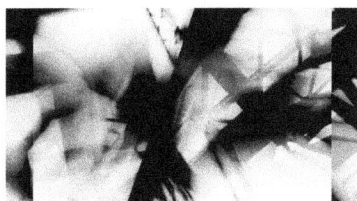
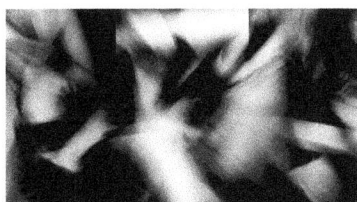
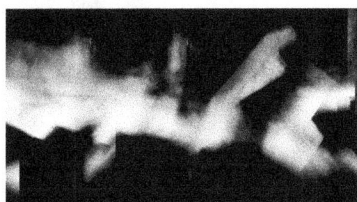

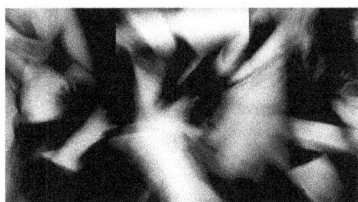

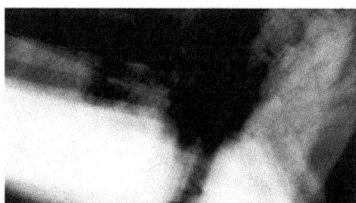

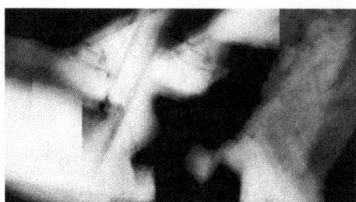

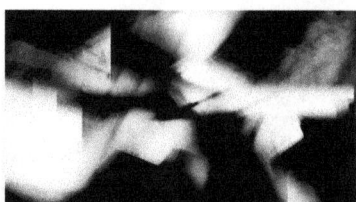

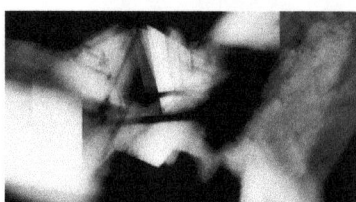
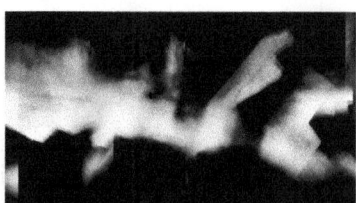
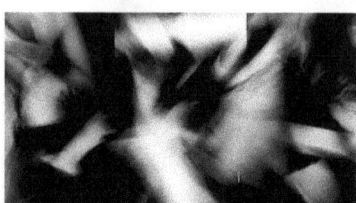

www.ingramcontent.com/pod-product-compliance
Lightning Source LLC
Chambersburg PA
CBHW072052040426
42447CB00012BB/3098